Math in the Garden

by Joanne Mattern

Red Chair Press   Egremont, Massachusetts

Look! Books are produced and published by Red Chair Press:

Red Chair Press LLC   PO Box 333   South Egremont, MA 01258-0333

 FREE Educator Guides at www.redchairpress.com/free-resources

Publisher's Cataloging-In-Publication Data
Names: Mattern, Joanne, 1963- author.
Title: Math in the garden / by Joanne Mattern.

Description: Egremont, Massachusetts : Red Chair Press, [2022] | Series:
    LOOK! books : Math and Me | Interest age level: 005-008. | Includes
    index and suggested resources for further reading. | Summary: "Math
    isn't just something you learn in school. It's an important part of the
    world around you. Young readers will use math skills such as measuring,
    counting, and adding as they plant flowers and vegetables in the
    garden"--Provided by publisher.

Identifiers: ISBN 9781643711317 (hardcover) | ISBN 9781643711379
    (softcover) | ISBN 9781643711430 (ePDF) | ISBN 9781643711492 (ePub 3
    S&L) | ISBN 9781643711553 (ePub 3 TR) | ISBN 9781643711614 (Kindle)

Subjects: LCSH: Mathematics--Juvenile literature. | Gardening--
    Mathematics--Juvenile literature. | CYAC: Mathematics. | Gardening--
    Mathematics.

Classification: LCC QA40.5 .M384 2022 (print) | LCC QA40.5 (ebook) | DDC
    510--dc23

Library of Congress Control Number: 2021945364

Photo credits: Cover, p. 1, 3, 5–11, 13–23: iStock; p. 4, 24: Shutterstock

Printed in United States of America
0422 1P CGF22

Table of Contents

In the Garden

You can do many things in the garden. We can see plants grow. Read on to see how we can use math in the garden.

Think ahead. Will a plant grow to be big or stay small?

Let's Start Digging!

You need to dig holes for your new plants. Each hole should be 8 inches deep. Find 8 inches on a ruler. Then dig your hole.

This **row** is 36 inches long.
Each plant should be six
inches apart. How many
plants can fit in this row?
Count by sixes to find out.

6 + 6 + 6 + 6 + 6 + 6 = 36

MATH FACT!

We can fit 6 plants in this row.

The seed **packet** says to plant three seeds in each hole. How many seeds will you need to fill six holes?

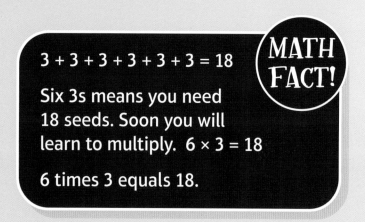

3 + 3 + 3 + 3 + 3 + 3 = 18

Six 3s means you need 18 seeds. Soon you will learn to multiply. 6 × 3 = 18

6 times 3 equals 18.

MATH FACT!

How Much Water?

The seeds are in the ground. Let's water them. Each hole needs 1 cup of water. How much water do you need to water all the plants?

1 + 1 + 1 + 1 + 1 + 1 = 6

You need 6 cups of water.

MATH FACT!

13

Big and Little

Here are some flowers we planted a few weeks ago. Let's measure them. The first plant is 6 inches tall. The second plant is 9 inches tall. Which plant is taller?

9 > 6

MATH FACT!

The 9-inch plant is taller. You say this as 9 is **greater** than 6.

Harvest Time

These tomato plants are ready to **harvest**. The first plant has 5 tomatoes. The second plant has 7 tomatoes. The third plant has 4 tomatoes. How many tomatoes do you have?

$5 + 7 + 4 = 16$.

You have 16 tomatoes altogether. Group these numbers to add:

$5 + 7 = 12$
$12 + 4 = 16$

MATH FACT!

Not all tomatoes are red when ready to pick. Some are yellow or orange and even green with stripes!

Each basket holds 8 tomatoes.
How many baskets do you need
to hold 16 tomatoes?
1, 2, 3, 4, 5, 6, 7, 8 tomatoes
fit in the first basket.
1, 2, 3, 4, 5, 6, 7, 8 tomatoes
fit in the second basket.
You need two baskets.

Let's pick the flowers in the garden. How many flowers do you have? Let's count.
1, 2, 3, 4, 5, 6, 7, 8, 9

No matter how many flowers you have, they can make someone smile.

A Great Garden

Making a garden is fun. You can use math in so many ways by counting, adding, and measuring!

Words to Know

harvest: to pick fruit and vegetables

packet: a small paper that holds tiny objects like seeds

row: things in a straight line

Learn More at the Library

Check out these books to learn more.

Levit, Joe. *Let's Explore Math (Bumba).* Lerner Publications, 2019.

Shah, Keiran. *Math in the Garden.* Gareth Stevens, 2017.

Steffora, Tracey. *Measuring in the Garden.* Heinemann-Raintree, 2011.

Index

About the Author

Joanne Mattern is the author of many books for children. She loves writing about sports, animals, and interesting people. Mattern lives in New York State with her family.